Life After Peak Oil

Life & Light

Copyright © 2014 Life & Light

All rights reserved.

ISBN: 9781494918521

Earth is in danger in many ways. This book focuses on the energy problem. Comfortable living from fossil fuels is soon coming to an end. What should humanity do? Let's take a closer look at the power of fossil fuels and the crisis that we are facing

CONTENTS

1 END OF FOSSIL FUEL CIVILIZATION 1

2 QUESTION OF ALTERNATIVE ENERGY 16

3 GLOBAL WARMING, WHAT IS THE CAUSE? 25

4 NUCLEAR ENERGY, CAN IT ENSURE OUR FUTURE? 37

5 QUEST FOR A NEW CIVILIZATION 52

1. END OF FOSSIL FUEL CIVILIZATION

Fossil Fuel and Our Lives

Humans have been using fossil fuels for a very long time, but now we are dependent. The word "Homo Oilicus" is a sarcastic reference to the modern humans who eat, use, and wear fossil fuels. The daily oil consumption in the world was approximately 86 million barrels in 2010 and is projected to be over 90 million barrels by 2015. This is equal to about 135 to 150 oil tankers emptied each day. We are "addicted to oil".

For modern humans, oil is almost as important as water, air and sunlight. Fossil fuels are energy itself. It has produced electricity, warmed buildings and allowed traffic systems to exist. Without energy, everything in our lives will not function. Civilization was able to develop so rapidly and vastly because of energy. Normally, people think of something huge when they hear the word "civilization". However, "civilization" can also be thought of as simply our normal daily lives. Today, our everyday lives and civilization are comfortably sustained by fossil fuels.

This calls for a review of our "oil addicted" lives. Daily life can largely be divided into food, clothing and shelter. Some people may have confusion about how oil and these life necessities are connected. Most people connect fossil fuels to industrial systems, but they are connected to many more aspects of life, almost all in fact.

First, let's look at clothing. Clothes act as basic protection for our body. Almost all other animals except for humans have fur and tough skin for self-protection. Throughout most of history, our clothes were completely based on and came from nature. Clothes were always hand-made. Therefore, it was impossible to mass-produce. The old 'self-sufficient' lifestyle was just enough and natural.

Our lives have changed a lot from those old days. How did the culture of modern clothing begin? As the oil civilization started, and industrial techniques were started, we began to gain access to comfortable, durable and economic clothing. Mass production allowed for less concern with wearing clothes for protection, and became more a method of expressing ourselves. Nowadays, we want and own more clothes than we need, in order to differentiate and express ourselves. This would not have been possible without oil and industrial processes. We should thank oil for how many clothes we have in our closet.

Oil also affects our dietary lives. Like clothing, for most of human history, diets were subsistent and self-sufficient. Small numbers of animals and plants were usually bred nearby for food. However, that dietary life began to change long ago. Globalization of the food industry started long before globalization became a trend. Large portions of the food consumed now is not produced locally, or even in the same country. The rest is imported from other countries where the food is mass-produced at a cheap price, and then transported with the help of oil. The way food is cheaply mass-produced is also deeply interconnected with oil. This is easily seen with the example of beef.

Most countries import beef from the United States, Canada and Australia. The reason beef from these countries is so economically competitive is not only because of the advantageous environment, but also because the cows are made fat rapidly by feeding them with grains and meat. Oil-based products fertilize many of these grains. The problem with this beef production method is that it disrupts the ecosystem and also consumes too much oil. In the U.S., it requires approximately one barrel of oil to produce one cow. After that, oil is needed to export the beef, and energy is required to keep the meat refrigerated the entire way. To sum up, the beef we eat requires large amounts of energy and resources. So, if oil prices increase, the importation of beef for all other countries becomes harder and more expensive. As the cost of oil and importation becomes higher, where a country cannot be self-sufficient, it will eventually lose access to those types of food.

Oil is also required for producing rice and grains. Fertilizers, pesticides and farming machines all require oil. Then, these harvests are all transported by using oil. In this context, modern farming is also called "oil farming". The saying "oil is food" is not too far off.

Lastly, let's look at our shelter. In this case, it is easier to notice the work of oil because we see energy being used; and everywhere energy is being used, oil is being used. From the building process, to heating and using electronics in the building, oil and its by-products are needed. Also, plastic products are abundant and almost all are made with oil. We are indeed living in oil shelters.

Although a shelter's role is simply to block the coldness and rain from the outside, shelters nowadays provide a lot more meaning. As individualism became stronger, people searched for personal meaning through their homes and individualized rooms. Also, rooms with special purposes appeared, such as dressing rooms and baby nurseries. This desire led people to prefer bigger houses. Bigger houses meant that more energy was needed to produce the structure, and then regulate temperatures, and of course more things were needed to fill the empty spaces. It could be said that unnecessary spaces are being filled with unnecessary things. We should be able to notice that oil has provided us with so much more than we can manage. Also, as houses became targets for investment, desire for houses grew extreme.

This is the foundation for a real problem. Fear and chaos are coming upon the oil-addicted people of the world. We are approaching the peak of oil production and consumption, which is the time when oil demand exceeds the supply.

The Perils of Peak Oil

Two synonyms for the word "peak" are "summit" and "climax". So, the phrase "peak oil" can be understood as the summit or the climax of oil production. However, anything that has reached its maximum is bound to fall. Therefore, the fact that oil production has reached its peak signifies the downfall of oil civilization and also is a notice of the vast changes to come.

What then is peak oil exactly? It does not refer to the point when all the oil on earth has been used up. It refers to the point when the oil demand exceeds the oil supply. To understand this, we must first take a look at the oil production process.

After oil is found, production starts with the use of wells to bring the oil from deep underground. When an oil well is first installed, oil is pumped up easily due to high pressures under the ground. The demand for oil led to more and more wells and a massive increase in oil production. Today, the production quota of OPEC, the world's largest oil-producing organization, is activated and the production rate is set according to the consumption rate. However, as more oil is pumped up, the pressure decreases and the viscosity of the oil increases and the emission speed of the oil inevitably decreases. These factors make production more expensive, energy intensive, and the resulting oil decreases. Eventually, injecting high-pressure water is needed to pump up the oil, but the production efficiency and quality of oil will continue to decrease. This method uses large quantities of water, which is also a dwindling precious resource. It also leads to the need to dig deeper and deeper into the earth, which can lead to the dangerous effects seen with the Deepwater Horizon disaster. Accordingly, the decrease of oil production is very natural.

The production of limited resources increases as consumption increases and it becomes a problem when the consumption rate exceeds the production rate. The oil peak refers to the point this happens and it is the fundamental reason for the crisis of our oil civilization. Yet, people start to get confused with the point of peak oil. Why is that?

Normally, the amount of oil produced from any single oil well takes the shape of a bell, known as a bell curve. After the maximum point of production, the amount of oil "gradually and steadily" decreases. Picture this same thing happening to all the wells on the planet. Peak oil can cause confusion in people's judgments if they do not understand. The gradual decrease allows for the illusion that there is no serious problem in living as we have "for the time being". However, can we live the same way without any problems after the oil peak?

We cannot be sure about the global oil-production bell-shaped graph. The graph is based on assumptions about known reserves and production and consumption rates. The fact that one day the consumption rate will exceed the production rate is absolute, but we cannot know whether the oil will decrease gradually and steadily after the oil peak. As mentioned earlier, oil production rates are set according to the consumption rate. The circumstances must change after the oil peak. Then, the consumption rate must be set according to the limited production rate.

A consumption quota will be needed instead of the original OPEC production quota. Therefore, the graph after the oil peak cannot be predicted. No one knows whether the graph will gradually decrease, drop suddenly, or be able to maintain the maximum rate for some time. It will depend on consumption rates, newly found oil reserves and the decision of powerful nations. What is important to understand is that the oil peak is imminent.

Researchers have brought up the oil peak problem for a long time. One famous example was King Hubbert, a geologist who worked at the Shell research lab.

Hubbert believed that the output of oil in a specific region could be calculated using the principle used in measuring the output of each oil well. In 1956, after thoroughly investigating the rate of the discovery of the oil wells, the production rate of oil and predicted oil deposits in the 48 U.S. states, he announced that the output of oil in the U.S. would reach its peak in 1970. This prediction came true and the U.S. became an oil importing country after the 1970s.

After 2000, the oil peak problem has been globally revived due to the work of Professor Kjell Aleklett and ASPO (The Association for the Study of Peak Oil & Gas). The work on peak oil is consistently being researched upon and discussed about in many different organizations. Each organization gives different analyses and forecasts about when oil will peak. Additionally, the oil-producing countries belonging to OPEC refuse to disclose their oil deposit quantities, which makes the exact calculation of the point of oil's peak that much harder. Despite the various forecasts and factors, most point to the years between 2010 and 2015 as the point of oil's peak.

On October 17, 2007, about five hundred oil experts participated in the World Oil Conference that was held in Houston, Texas. Many there had the opinion that peak oil had already passed in the year 2006. The general opinion was that the oil peak would arrive by 2011. Professor Lester Brown stated in his book, *Plan B 3.0*, that many geologists assumed the year 2010 was the point of peak oil. Also, he warned that the price of oil will skyrocket and the competition for energy will be fierce.

Although the opinions of the experts vary, they all agree that the oil peak will come in less than twenty years. Many believe it has already passed. Considering that the price of oil reached 147 dollars per barrel in the year 2008, scholars around the world are considering the possibility that the oil peak has already arrived. Actually, many non-OPEC oil-producing countries passed peak oil around the year 2000.

Of course, not everyone believes these opinions. The opponents often claim that the premise of the peak oil theories, assuming that the total amount of oil deposits is a fixed value, is wrong. They argue that the oil deposit can vary due to the oil price and the development of mining technology. They claim that if the oil prices rise, the oil wells that have been stopped due to high costs will re-open, and technological development will allow access to oil in deeper and deeper places. Taking these claims into account, they estimate the world oil deposits are one trillion barrels higher than standard estimates and that peak oil will not arrive for another twenty to thirty years.

Proponents of peak oil theories rebut that most of the places on earth have already been explored and that there can't be one trillion barrels of oil hiding. Even if there were a trillion more barrels in the world, the peak would only be pushed about twenty more years in the future. It is not very far in the future and the same examination of oil's central role in our lives, and perhaps changing our lives to better fit the natural system, is just as warranted.

It is also important to understand that the proponents of peak oil are mostly geologists who have worked in the oil industry and have practical experience, while the opponents are mostly economists. The best energy geologists are warning that the oil peak has already passed or is to come in a few short years. As mentioned earlier, those who disagree are limited to claiming that we have only ten to twenty more years.

Peak oil is a major problem because it is the start of a downhill slide. As soon as the downhill starts, the production rate fails to catch up with the consumption rate. The peak might come much earlier because of the recent increases of oil consumption in China and India. As their economies grow, they will consume more energy and the peak will arrive earlier. If that happens, all countries will compete for more oil; and eventually, under the lead of the most powerful nations, oil will be distributed according to a quota system. Then, oil prices will rapidly skyrocket and the world will be forced to head towards structural reformation. In other words, the world, which was even shaken by a simple oil shock in the 1970s, will fall into chaos, as countries will attempt to steal other countries' oil by force. That is, if oil remains central to our lives.

It is important to understand that the oil shock was something "simple". The oil shock was a temporary supply and demand problem caused by OPEC regulations, not a fundamental global oil shortage problem. Those problems were all solvable. However, the future oil shock after the oil peak will be unsolvable. It is not a problem of adjusting the supply and demand, but a problem of having no oil to supply. Now we come to the situation where we cannot adjust the amount of oil supply. The people who control the world might be hiding the peak of oil production on purpose because there is no solution and they can manipulate its price, but it will still come either way.

Yet, the market economy seems to be aware of the situation. The fluctuation of oil prices looks unusual and is influenced by speculation. Look at the global economy of the years 2008 and 2009. In the 2000s, the global economy has grown largely around the U.S. and the BRIC countries (Brazil, Russia, India, China). The rate of industrial operation increased worldwide and the growth led to an increase in both housing demand and supply. The increase in the housing supply meant that the production of building materials and electronics, which are all produced by using oil, also reached their climax with this demand.

During that period, the demand for oil naturally increased and the Bush administration asked Saudi Arabia for increased production. However, the request was refused. The May 18, 2008 *Wall Street Journal* report handled this refusal in detail. Saudi Arabia, one of the biggest oil producing countries, either was not able to produce more oil or was trying to sell the limited amount of oil at higher prices. This led to a rise in oil prices and this eventually affected the housing economy.

Having reached its limits, high house prices severely dropped after the bankruptcy of the Lehman Brothers in 2008. Many other financial institutions, which based their business management on mere numbers, began to also go bankrupt, and this led to a drastic disruption of the global economy. This led to limitations on housing prices, especially high-rise housing, due to the concurrent rise of oil prices. The banks judged that further increases in housing prices were unlikely and started to withdraw mortgage interest loans. When these facts were combined with a flood of houses for sale, the overall value of many homes dropped dramatically.

Following the decrease in mortgage values, more financial institutions went bankrupt and the housing economy went into a depression; and this, in turn, affected the oil supply and demand by lowering the demand for furniture and electronics. This time, the problem was that the decline in oil demand meant that there was a shortage of space to store the surplus oil. Therefore, oil was urgently sold and caused the oil prices to rapidly drop towards the end of 2008.

As energy costs became cheaper, costs of production decreased and allowed the promotion of industrial actions and setting lower prices for goods. But, the oil price has climbed back up and is now heading for an even higher peak. Although the greedy financial companies are mainly the ones to blame for the global economic crisis, the short construction boom increased oil consumption to over 86 million barrels per day and the fact that the production rate could not adjust to meet the need shows that oil production has reached its peak. The market economy is being controlled by the oil price. What is even more shocking is that the market economy, which is being controlled by the oil price, is controlling our lives. This means that oil is dominating our lives. We are living in a time in which oil is dominant over our daily life, economic activity and global economy.

Life After Peak Oil

2. QUESTION OF ALTERNATIVE ENERGY

The previous chapter established that oil is maintaining our everyday lives. This oil-dependent modern life sometimes seems unhappy compared to the old self-sufficient life. Everything that we use, ranging from food to our houses, cannot exist without the help of oil.

Having realized the crisis, it is important to quickly find an appropriate substitution for oil. Our problem will be largely solved if such an alternative source is found. However, scientists have failed to produce outstanding results in researching alternatives like solar and wind power over the past thirty to forty years.

The Reality of Alternative Energy

Many scholars and books mention the importance of alternative energy as a solution to the crisis that will arrive once fossil fuel is depleted. Alternative energy, as is mentioned in its name, is an energy that substitutes the role of fossil fuels completely. Most people do not worry much about the depletion of fossil fuels because they believe the scientist's claims that solar and wind power will be able to replace fossil fuel. However, there is a big problem. Alternative energy must be able to substitute one hundred percent of oil.

Current alternative energy technology will never be able to replace oil. The reason is that alternative energy can only be produced with the help of oil. We are claiming that solar cells and wind power generators will replace oil, despite the fact that we need oil to produce and install the cells and generators.

Consider how much time we spent in a vehicle and how many vehicles go past us. Can vehicles function without oil? In science fiction movies, airships fly instead of cars. And we all vaguely dream that it will become a reality. A world with automatic airships that activate with one button is a dreamland that is both comfortable and fast-moving.

However, what is the reality? How are we to travel without oil? Many people think of electric cars. Electric cars have been slow to appear, but the intermediate level, hybrid car use is slowly growing. Still, it seems hard to believe that electric cars will be able to perfectly substitute the roles of the original cars. The main characteristic of electric cars is that they use electricity instead of oil, but the electricity must be produced from something other than oil, like nuclear energy. This also has plenty of problems and will be discussed in a later chapter.

Since we cannot always have an electrical cord connected to the car, a storage battery is needed. The problem with storage batteries is that they become too big and heavy as the capacity increases, restricting usage to small cars, and it takes too long to charge, making it incompatible for 'refueling' on long distance travels. Also, the efficiency of the storage battery decreases as time passes and disposing used-up batteries is not easy because of the pollution it causes. Current battery technology also uses rare earth metals that are extremely difficult and energy intensive to produce.

It must be realized that oil is currently needed to produce electricity, biofuels and hydrogen. To mass-produce the raw material for biofuel, such as corn or sugar cane, petrochemicals and agricultural machines are needed. Therefore, the paradox is the fact that alternative energy, unlike its name, cannot perfectly be an alternative for oil. Oil is still needed to run the facilities required to convert raw forms of energy into energy we can use.

The problem must be approached at the fundamental level. This is the only way to find an answer. The so-called alternative energy is currently only an extension of oil. It only allows us to use oil a few years longer than if we didn't use it. It is no different than attaching a respirator to a dying patient.

Let's look more closely at alternative energy, which we thought was a certain replacement for the future and a solution to our oil dependence.

No Answer from Solar and Wind Power

Solar and wind power are mentioned as excellent alternative energies. But, basic problems are found with these two alternative energies also.

Let's consider solar energy. First, fossil fuel energy is needed to manufacture the solar cells. 2,000 watts/meter^2 of energy falls upon the solar cell panel, but only 100 W/m^2 is converted to electricity. This means it is only five percent efficient. On top of that, it can only be operated for an average of four hours a day. Therefore, with an entire roof filled with solar cells, we can only barely power our home, but we cannot air-condition or heat the house. Even though solar cell efficiency is improving, these basic problems will remain.

Wind power is relatively widely used in Europe, and is rapidly emerging as the leading future energy source. Wind power has the largest annual increase of usage among all the alternative energy sources. However, the scale of the wind power industry that is needed to reach an appropriate level presents several problems. First, wind power is not unlimited. There are only so many places where high quality wind constantly blows. According to experts, high quality wind refers to wind with uniform strength and direction. For that reason, it is hard to build wind generators in mountainous areas, where the wind blows irregularly, which forces the generators to vibrate dangerously. It also isn't easy to build durable wind generators because they are colossal structures and require too many special materials and energy to produce. Therefore, most wind power generation complexes are located at sea-level coasts, but this also creates problems of people being forced to abandon the fishing industry due to the blockage of ships to install power lines for generators.

The persisting period for high-performance wind power generators is about twenty years, and they require constant maintenance the entire time. Also, since the wind generators are usually built far away from the main electricity generating facilities, it is very costly to build power lines to connect to the existing grid. Therefore, there is still much to consider if we are to use wind power as a main source of energy.

As we saw earlier, the pros and cons of solar and wind power are generally well known. So, we will take a slightly different approach and look at efficiency. Let's compare size of the solar and wind power needed to produce the same amount of energy as one nuclear generator. If they are to truly be the alternative energy, their size and output must be realistic. The capacity of each nuclear generator is about 1,000 megawatts; how many solar cell panels and wind generators are need to generate this much energy?

First, solar power is created from the nuclear fusion reaction between hydrogen molecules and the amount of solar energy that falls upon Earth. The energy the sun gives to the Earth in one hour is equal to the amount of energy that the entire planet uses for two years. Although solar energy is vast and everlasting, it is hard to use efficiently. This is because the energy density is too low and is largely influenced by temperature changes during the day.

How solar power is collected, converted and stored determines its economic feasibility. Even though we have improved solar collectors and solar cells, their efficiency is still a big problem. Currently, connecting several modules to make a 1 m^2 solar cell can generate about 100W. Based on this, to make a solar cell panel equivalent to a 1,000MW nuclear power plant, we need 10 million m^2 and a sunny area of twice the size in order to install the solar cell panels. Also, the operation time of a solar cell is typically about five hours, while nuclear power plants can be run for twenty hours. This means that the efficiency of the solar cells must be increased by four to five times. Huge amounts of land and solar cell panels are need to equal the capability of a single nuclear power plant.

Wind power also has the same problem. Sunlight does not hit the surface of the Earth uniformly. Therefore, the surface temperature differs with region and the land and sea are heated differently. The difference in temperature creates wind. Wind power uses this wind to run the turbines and obtain electricity. The capacity of wind turbines ranges from 10W to 5MW. Typical large turbines have blade diameters of seventy to eighty meters and the height of the column is much bigger. Comparing a 1,000 MW nuclear power plant with a high-capacity 2 MW wind turbine still requires 500 wind turbines. Since the wind turbines have a height of 80 to 100 meters and blade diameters of 70 to 80 meters, the installation area for each turbine must be at least 200 meters apart. Also, if 100 units are installed in each of five rows, the total area is 20,000,000 m^2. The land needed for installing the required wind turbines is equivalent to that for installing solar cells.

The use of fossil fuels for mining, transporting, manufacturing the materials required for solar panels or turbines, and installing the generators is inevitable. In other words, it cannot function without the support of fossil fuel based industries. On top of that, the products only last twenty to thirty years, which makes it an inefficient, resource-wasting industry with many limitations.

Also, these new energy resources are more damaging to the environment than was originally thought. For example, excluding the ten to twenty percent of the solar energy that is changed to electricity in solar panels, eighty to ninety percent of the remaining radiant heat causes a surrounding heat island. With wind turbines, the ecosystem is severely affected by the noise and blockage of wind. When used on a large scale, both technologies disrupt wildlife and ecosystems.

If oil were to disappear tomorrow, manufacturing alternative energy facilities becomes impossible. In other words, it is realistically impossible to live with alternative energy once oil is depleted. Alternative energy is merely an auxiliary material of fossil fuels.

Each day, the human population is increasing by 250,000 people, 1,000 hectares of tropical rain forests are being destroyed, and three species are being led to extinction. We are destroying many things as we run towards the peak. The peak will arrive earlier if we do not reduce our energy consumption. What should we do? Is nuclear energy the key to all this? Nuclear energy is considered eco-friendly because it does not emit CO_2 during the nuclear fission reaction. Therefore, it is important to, first, look at the relationship between CO_2 and global warming. Then, we will see if nuclear energy is indeed economical and clean, and whether it is abundant enough to completely substitute the role of fossil fuel.

3. GLOBAL WARMING, WHAT IS THE CAUSE?

Quit Blaming CO_2

The U.N. has recently announced that Earth is under severe global warming, and its main cause is CO_2 produced from fossil fuel. However, is the global crisis really caused by CO_2? Will all organisms on Earth really be wiped out because of CO_2? As mentioned in the beginning of the book, it is important to question and research what is considered fact and find the real underlying crisis. Therefore, although it seems there is no room to have space for doubt, let's look at the relationship between global warming and CO_2.

The Intergovernmental Panel on Climate Change,

IPCC, an affiliated organization of the U.N., announced its initial global warming report in 2007. This report took six years to complete and had contributors from 130 countries and 2,500 experts. It predicted that the temperature of Earth will rise by a maximum of 4°C in the next 20 year span, pointing out the use of fossil fuel as the main cause for this to occur. The fifth report, put out in 2013, has reduced that alarmist estimate to about one degree Fahrenheit. According to the IPCC, the main gases that cause global warming are CO_2, methane, PFC, nitrous oxide, etc., and CO_2 is known to cause fifty percent of global warming. CO_2 is produced from all combustion reactions of coal or oil.

According to the Kyoto Protocol signed in 1997, advanced countries must reduce greenhouse gas emissions to an average of 5.2 percent lower than the emission level in the year 1990. However, there are different opinions among advanced and developing countries about the amount to reduce, schedule, and who is to participate. On top of that, despite having 28 percent of emissions, the U.S. withdrew from the group in March 2001, in order to protect its domestic industries.

During the G8 summit of 2008, it was agreed to reduce the emission of greenhouse gases by half of that year's totals by 2050. Also, measures were presented for developing countries that do not want their economic growth hindered by the economic constraints, and promises to increase investment costs for clean techniques were made.

CO_2 is considered the main cause of global warming and reducing CO_2 is viewed as the main point in solving the problem of global warming. Environmentalists and summits around the world talk about reducing CO_2. Why is CO_2 pointed out as the main cause of global warming? There are two main reasons.

First, CO_2 re-radiates light in the infrared wavelength range and acts as a film covering the Earth. This CO_2 film prevents the light energy from the Sun from being released into space and this characteristic is considered key. The radiant energy from the Earth is trapped in the atmosphere and causes the greenhouse effect.

Secondly, scientists researching global warming have gathered climate data back 650,000 years ago from ice samples in the South Pole. They found that during interglacial periods, the concentration of CO_2 increased. From this conclusion, the amount of CO_2 in the atmosphere has become the major standard in determining global warming.

But, there are at least two reasons to doubt this

conclusion. First, in order to discuss the greenhouse effect of CO_2, the amount of energy stored in the infrared wavelength range must be considered.

Infrared light refers to light with a wavelength ranging of 10^3 nm ~ 10^6 nm; considering the whole range of light, which is from radio rays with wavelengths of 10^3 m to gamma rays at 10^{-5} nm, infrared refers to only a sliver of the light spectrum.

According to Planck's quantum theory, the energy of light is '$E=hC/\lambda$' (E: energy, h: Planck's constant, C: speed of light, λ: wavelength of light), where energy is inversely proportional to the wavelength. Comparing the infrared wavelengths with visible light, which is 400nm~700nm, infrared light has relatively small energy, less than one-tenth of the energy of visible light. This raises a number of questions.

First, the infrared light that CO_2 radiates has an energy that is only ten percent of visible light. This means the amount of infrared radiation and the ratio compared with visible light should be examined, when the sun's radiation energy enters Earth and is re-emitted by the Earth. The existing research only considers the total amount of energy of sunlight and neglects the energy calculation according to each wavelength. Are they assuming that all solar energy that comes to Earth is emitted only as infrared light? Surely not, because if that is the case, we will only be able to see Earth with an infrared camera, but we know from the satellite pictures that there is a lot of visible light. It is important to know how much of the Earth's energy from the sun is infrared.

Secondly, the atmosphere contains less than .04 percent of CO_2. This means that only four molecules are CO_2 among ten thousand air molecules. Are these four molecules the cause for the heating of Earth? Also, since the overall temperatures of all the molecules are the same, these four molecules would have to affect the ten thousand molecules in the same way. Is nitrogen and oxygen, which makes up 99 percent of the atmosphere, not influenced by any radiation re-emitted by the Earth? Yet, in reality, the temperature of nitrogen and oxygen also increase when exposed to light or energy. Rethinking, the former global warming theory, which states that .04 percent of air molecules are the cause of global warming, is necessary. Careful and rational experiments are needed to prove that it has a superior temperature-rising ability than the other 99 percent of air molecules of nitrogen and oxygen.

Third, they say that the CO_2 gas repeatedly emits infrared light and this has the effect of one thousand to ten thousand times the enthalpy of the formation of CO_2 (393 KJ/mol). The same people claim that global warming is accelerated due to repeatedly radiated earth radiation, instead of new energy that is generated by the sun. Is this claim correct?

Fourth, if CO_2 reemission of infrared light is the cause for global warming, we must also consider the infrared light that is blocked and reemitted from the sunlight. The infrared in sunlight and the radiation of infrared from Earth must be compared and then used as a basis for judging CO_2 impact.

Fifth, the explanations of CO_2 gas experiments argue only that it reemits infrared light. Is it completely unaffected by the visible light range?

Environmental Experiment

A simple experiment was conducted to investigate the radiant effect of CO_2. First, three transparent glass boxes were made. Each box was divided into top and bottom parts and equipped with sensors that recorded the concentration and temperature change. Also, white Styrofoam covers were installed on the bottom and sides of the box so that sunlight only entered through the top. Next, all of the boxes were filled with ground air and the temperature changes were measured several times to calculate the site errors and measurement errors. Then, with the exception of one reference box, the top of one box was filled with CO_2 and the top of the third box was filled with Helium at a concentration higher than eighty percent. Then, the change in concentration and temperature was measured repeatedly for a period of time.

The results showed that during the day, the box with the higher concentration of CO_2 had higher temperature increases in the top CO_2 part of the box and lower temperature increases in the bottom air box compared to the reference box. During the day or on cloudy days, the box with the higher concentration of Helium resulted in lower temperature in the top part of the box. The bottom air part of the box was hotter than the reference box during the morning but gradually cooled after noon. There were no differences found between the boxes during the night. The concentration of CO_2 and Helium gradually decreased and after ten days, it decreased to below twenty percent and approached normal atmosphere temperatures. Overall, the temperature of the bottom part of each box was twice as high as the top part.

The experiment showed that temperature is influenced by sunlight and when it was cloudy or dark, the boxes had no temperature difference. This shows that gas temperature is closely related with light, and temperature change occurred due to the radiant characteristic of each gas. Particularly important in this experiment was the fact that the CO_2 box was hotter than the ground air box, which means that the radiant effect of CO_2 is better than that of air. The fact that the bottom part of the CO_2 box had lower temperatures than the bottom part of the ground air box is important. This experiment shows that the surface temperature could, in fact, get lower because CO_2 blocks sunlight, which would put more of the focus on the cooling tendency of CO_2.

The temperature of the Helium box was lower than the reference air box, but the bottom part of the Helium box was hotter during the morning because there was less cover against the sunlight, but then eventually cooled due to the cooler top part of the box. Alternatively, although the temperature of the top part of the CO_2 box was high, the temperature of the bottom part decreased, hinting at the validity of the cooling effect of CO_2.

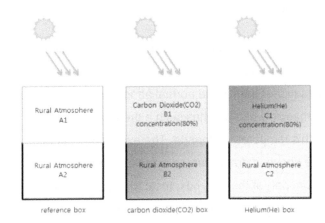

reference box   carbon dioxide(CO2) box   Helium(He) box

One could argue that water vapor could be a greenhouse gas as well, but clouds, which are a big group of water vapors, actually block the sunlight and therefore has a more cooling effect.

What might be the reason for this cooling effect? The gas molecules with greater mass probably have better radiant effect, and thinking logically, heavier materials get hotter faster. The existing theory of CO_2 as a greenhouse gas is a result of only considering the infrared radiation from the Earth's surface, while neglecting the net sunlight reaching Earth.

This simple experiment was not perfect. However, it is enough to reconsider the narrative that a CO_2 caused global warming crisis is about to come. The only way to proceed is to find the accurate causes and make precise countermeasures.

It is important to emphasize the fact that our experiment used a CO_2 concentration two thousand times higher than in the normal atmosphere, in order to emphasize whether it has such a great impact on the Earth's temperature change. The experiment should have shown a drastic increase in temperature in the bottom part of the CO_2 box. Instead, it gave a completely different result and showed cooler temperatures because the CO_2 blocked some of the sun's energy to the bottom part of the box. If CO_2 has a much greater warming effect than other gases, shouldn't it have been shown by the temperature of the box? Because all other conditions including re-radiation were the same in this experiment, if the box below the CO_2 box is cooler, then it shows that CO_2 is cooling indeed.

Let's relook at the glacial data. The data from greenhouse gases measured from ice samples in the South Pole must be examined considering the specific situation of that time. For example, if there was a volcanic eruption or a massive forest fire that human efforts never extinguished, naturally, there would have been more emissions of CO_2 than in other years. Rather than thinking that CO_2 influenced, or caused, global warming, couldn't it also be possible that the heat from these fires represents the interglacial period and higher

CO_2 is a result of it? Also, when large volcanic eruptions go off, the planet experiences cooling because less of the sun's energy reaches earth due to the volcanic ash in the atmosphere, despite higher CO_2 levels.

The two main reasons we are given for why CO_2 is the main reason for global warming, infrared radiation and glacial data, are not completely trustworthy. Of course, the recent effort to reduce CO_2 may have helped to prevent global warming. This is because the effort led to reducing the use of the limited fossil fuels and this means less energy was released.

What is most important is the fact that the Earth is getting warmer is true, and the current situation, where we are focusing on a minor gas, CO_2, instead of looking for the essence of the problem is worrying. If CO_2 is not the main cause of global warming, worldwide efforts of dumping CO_2 below the ocean and scientific efforts to separate O_2 to effectively burn fossil fuel are useless because they focus only on CO_2. It is also being used to create carbon-taxing laws. This diversion might even make global warming worse.

CO_2 is a gas that has always been on Earth and is vital to the functioning of all plants and all organisms that are dependent on plants. In other words, it is also possible to see that more CO_2 means a better environment for organisms. Should we blame such a gas as the main cause of global warming, turning it into humanity's biggest enemy? Instead, we must look back at our selves. When a child has a fever, it would be harmful to the child if we act only based upon our interests without considering the actual cause of the fever. While the Earth is getting warmer, we are focusing on a gas and not thinking much about the heat we release every day. Besides, according to the experiment, the gas blocked more of the heat than it stored of the emitted heat. Isn't this strange? Having to contradict existing ideas usually does seem strange. What is scary is that no one has thought the existing ideas were strange, at least those given time in major media, and the general public has unconditionally followed them.

We must not to be afraid to question 'commonsense'. Avoiding questioning is the biggest obstacle to our potential of development, and is the cause of blindness and blind faith and of reduction of the width and depth of thinking. Even the slightest glance at human history shows how eliminating questions has allowed easy control of the public. It is important to know the real causes of global warming accurately. We can find that cause if we make an active and rational effort.

4. NUCLEAR ENERGY, CAN IT ENSURE OUR FUTURE?

As the previous chapter has shown, CO_2 is only a mark of energy consuming activity and cannot be the main cause of global warming. However, what is missed is that the main cause of global warming is energy consuming activities of humans.

Why try to control our lives by trying to reduce CO_2, if CO_2 is not the problem? It is important to understand that the final goal is not 'reducing CO_2'. We must realize that all our efforts are to 'reduce heat itself'. If we continue with our current energy consuming activities, the Earth will never be free from global warming. Also, focusing only on the result without controlling our consumptive desires will only incite the use of nuclear energy. But, if nuclear energy ends up having a much bigger problem than we thought, and it is only just another form of threat to Earth, what are we going to do?

Many people believe that nuclear energy is the only way to save the earth, now. They firmly believe that nuclear energy can solve the depletion of fossil fuel and the limits of alternative energy. This is not the only reason nuclear energy is being praised. It is also under the spotlight because of the fact that nuclear energy supposedly does not affect global warming. So, will nuclear energy be able to solve the depletion of fossil fuel and global warming? Let's take a closer look at the relationship between earth and nuclear energy.

Nuclear Renaissance?

On May 24, 2004, the British newspaper *Independent* published an article titled, "Nuclear power is the only green solution". The article was written by James Lovelock, who has continuously warned about greenhouse gases since the beginning of the 1970s using the Gaia hypothesis. Lovelock mainly talked about the importance of expanding nuclear power generation to stop global warming. Some environmentalists began to agree with James Lovelock's opinion. Although the sudden change in attitude of the anti-nuclear environmentalists is slightly weird, it shows that the sense of crisis from the depletion of fossil fuel is big. Nuclear energy emerged as a solution because there are no other suitable alternatives. It also drew people's attention because it does not produce CO_2, which is considered the main cause of global warming.

Nuclear energy became popular after the U.S. president Eisenhower advocated 'peaceful use of nuclear energy' at the U.N. conference on December 8, 1953 and reached its peak in the 1970s. Then, after the Three Mile Island accident in 1979 and the Chernobyl disaster in 1986, the safety problems were brought up and some countries adopted 'anti-nuclear policies' and the U.S. stopped producing more plants. The controversy around nuclear energy became more intense, and countries started to turn towards developing reusable energy, such as solar and wind energy.

Fueled by the idea that we can freely use the infinite amount of energy nature has, many countries began pouring huge amounts of money and manpower into developing reusable energy. However, the result was unexpected. The experts were slowly becoming convinced that it is limited to being a supplementary energy source because it is hard to mass produce electricity and not economical. Meanwhile, the rise of oil prices, sudden rise in energy consumption in developing countries like China and India, and resource hegemonies have forced the world to turn its eyes towards mass-producible and economical nuclear energy. The supposed nuclear renaissance began in the mid-2000s.

The U.S. restarted building nuclear plants for the first time in over thirty years, and announced and modified policies in order to expand the existing nuclear plant sites. Dormant European nuclear plant construction also woke up, starting with Finland's nuclear plant construction. Due to the energy law passed in June 2004, France planned to substitute 19 out of 59 nuclear plants to a new nuclear reactor model by 2020. Bulgaria, Romania, Ukraine and several other countries began planning or constructing nuclear plants. Even the anti-nuclear policy leader, Germany, also slowly changed its opinion in 2010 towards using nuclear energy after gas supplies from Russia became no longer stable.

Rapidly developing China and India are also building nuclear plants. China already has thirteen plants working and 27 in construction. They are planning to build fifty more by 2020 and activate 110 by 2030. Also, India has twenty plants working and five in construction. They are planning to build eighteen more by 2020 and forty more plants by 2030. Japan had 51 nuclear plants already working with 14 more to be made by 2020.

The International Atomic Energy Agency predicted that 659 nuclear plants would be operating by 2020. Adding the 326 plants that are being planned, the number of nuclear plants by 2030 was expected to double.

Some Middle East countries, such as UAE and Turkey, also began building nuclear plants in 2010. The fact that these oil-producing countries are acting in such a way shows that they are sensing an oil crisis as well. These anxieties are directly connected to the act of trying to substitute fossil fuels with nuclear energy.

The apparent strength of nuclear energy coincides too well with the current situation: countries are fighting over resources because of the depletion of energy resources due to unrestricted fossil fuel abuse. Also, the fact that nuclear energy does not emit CO_2 allowed it to be marketed as eco-friendly and made it even more powerful. Now, nuclear businesses, governments and even some environmentalists are advertising nuclear energy as 'economical, clean and eco-friendly energy' and the world has gradually adapted to such trends.

However, a number of factors has made the 'nuclear renaissance' already lose its luster less than ten years later, and many of the proposed plants are being cancelled. First, nuclear power plants have long been uneconomical and the private investing sector won't touch it. With the economic recession beginning in 2008 and without a global carbon-tax in place, the funding of nuclear plants is not economically viable.

Perhaps the most visible and long-lasting factor ending the nuclear renaissance before it began was the Fukushima Daiichi nuclear power plant disaster that occurred on March 11, 2011. After a massive earthquake and tsunami hit Japan, three of the six reactors at the Fukushima plant released large amounts of radiation after they overheated and exploded. The radiation release was estimated to be one hundred times worse than Chernobyl. What is even worse is that the plant has been forced to use massive amounts of water to cool the fuel cores. It is estimated that 300 tons of highly radioactive water is being released into the Pacific Ocean every day since the disaster and will continue until at least 2015. It could take decades to fully stop the release of radiation. This is wrecking untold damage to the ocean and the world. Untold, because those running the plant and the Japanese and American governments have a troubling history of lying about this event, the cleanup, and its environmental damage.

Two months after the disaster, Germany announced it would abandon all nuclear power by 2022 and shut down seven of their oldest power plants, which supplied almost a quarter of their electricity. They will shut down 17 more plants in the process. Japan also dropped its plans to build 14 more nuclear reactors. Both are moving towards solar energy to make up for lost power. The U.S. still currently has five plants being built, but others are being taken off-line and several other planned projects have been canceled. The U.S. currently has 100 nuclear power plants operating.

While the IAEA predicted 650 nuclear plants by 2020 only a few years ago, the actual number will be much lower. In 2013, there were 434 nuclear plants in operation and 69 under construction. This means the actual 2020 number will be around 500, but as countries continue to reconsider nuclear energy in light of the Fukushima disaster, the number will likely be lower. Additionally, many of these operating plants were built in the 1970s and will be about fifty years old by 2020 and in desperate need of renovating and risk an accident even worse than Fukushima. The renaissance was an illusion and unviable.

Limited Resource, Impossible Technology

The advocates of nuclear energy argue that its biggest advantage is a massive amount of energy from a small amount of fuel. The energy from nuclear fission and fusion is incomparably greater than the chemical energy stored in fossil fuels. This has led people to believe that we have abundant nuclear resources because we can gain a large amount of electrical energy from a small amount of uranium. However, most nuclear plants use uranium-235 to get nuclear fission energy and only 0.7 percent of natural uranium is uranium-235. If we trace the consumption trend, the time limit of using uranium-235 is similar to the fossil fuel depletion time. We would quickly have the same 'peak uranium-235' problem that we currently have with oil. However, by using fast-breeder reactors, which recycles the used nuclear fuel, we currently have an abundant supply of fuel, but this also produces many problems.

Nuclear fuels have an extremely high energy density, but they must first be processed before they can be used in reactors. The process of mining and processing uranium also produces much more cost and pollution than when mining fossil fuels, which are already significant. The cost of producing nuclear energy usually refers only to the fuel costs, maintenance costs and operating costs, including labor costs. In other words, the reactor shutdown costs, waste costs and waste storage costs are not included. On top of that, as became obvious with Fukushima, risk costs from accidents, high-level radioactive waste storage and disposal costs, and the pollution costs to the sea are completely excluded. If all of these costs are added, nuclear energy is likely the most costly and dangerous energy source.

To reiterate, the uranium itself is limited in amount, and to recycle the fuel, we must use plutonium. Plutonium is one of the world's deadliest substances, as less than five pounds is enough to poison all life on Earth. It is currently widely used for making nuclear bombs and certain types of nuclear reactors. Both uranium-235 and plutonium are hopelessly piling up as waste that will be radioactive for thousands of years.

Although energy generated from nuclear fusion is enormous, nuclear fusion reactors are in their infancy. According to the plan by ITER (International Thermonuclear Experimental Reactor), the experimental reactor will be made by 2020, and if it is considered usable, the nuclear fusion reactor will be made around 2040 or 2050. The main problems with nuclear fusion reactors are almost as enormous as the energy it could provide. The first and biggest problem is maintaining an energy level of over 100 million °C; the next problem is utilizing this temperature in the reactor. Considering that most materials on Earth melt below 2,000°C, no material is known that can withstand and contain 100 million °C. There is no practical material that can withstand such temperature. Currently, they are experimenting with a method where they use electromagnetic fields to float the interior materials (Tokamak), but this is still a long way from being practical. We do not have time to hope fusion can be commercialized by 2050.

We must rethink the idea that nuclear energy is abundant. Using plutonium as a nuclear fuel is technologically experimental and operating a plutonium plant, which has 10 times more uranium-235 than the usual nuclear plant, is completely different. When there is an accident, the risks are unimaginable.

The Severity of High-Level Nuclear Waste

High-level nuclear waste has been a known problem for a long time. However, as global warming emerged as a tangible problem, our attention was drawn away from this still unsolved problem, and many people have rashly started to support nuclear energy. It is important to look at nuclear power's long-existing second problem: high-level nuclear waste.

High-level nuclear waste refers to the highly radioactive materials that come out when nuclear fuels undergo fission. Radioactive elements, such as cesium, strontium, neptunium and plutonium, are some examples. There is a great problem in disposing high-level nuclear waste. It must first be enriched and solidified into a stable form, then temporarily stored in a stainless steel canister for about fifty years until its heat emission has reduced. Lastly, it is buried hundreds of meters underground. This means space is needed to store the high-level nuclear waste that is protected for thousands of years.

A single nuclear power plant discharges about one thousand tons of direct waste per year. This large amount of nuclear waste will continue to be a risk for millennia. These figures do not even include the one hundred thousand tons of uranium waste rock that is discharged annually in the production of fuel-grade uranium.

Would building a permanent storage site solve the nuclear waste problem? This belief is quite misguided. Although we have been developing and using nuclear energy since the 1950s, a permanent nuclear waste storage site has not been built. The reason is that the logistics to build an appropriate location that can safely store huge amounts of radioactive materials for tens of thousands of years is impossible. Despite having 434 nuclear power plants around the world, there is still no secure place and plan to store the nuclear waste anywhere. None of these countries has an ultimate nuclear waste disposal method. Although the nuclear reactor itself is managed to a certain extent, the high-level nuclear waste is simply gathered in one place without any preparation in case of a fatal accident. Considering the influences of high-level nuclear waste, the danger is high.

When high-level nuclear wastes are buried hundreds of meters underground, the ocean floor is considered an appropriate place. However, accumulated high-level nuclear waste produces heat and radioactivity. When high-level nuclear wastes are gathered in one place, it has a high probability to heat each other up, melt its containment reservoir, and even explode. Think of the danger of placing candles near combustibles, except the danger is exponentially higher.

Also, no water can ever leak into the reservoir. Water is sensitive to heat and would cause a great expansion pressure, causing a ground failure and radioactive leak. Once the storage space begins to melt and dismantle, the entire region nearby becomes extremely hazardous. A leak in a high-level nuclear waste storage facility can be more even more severe than a leak in a nuclear power plant. It will continue to emit fatal radiation for tens of thousands of years and the Earth could become uninhabitable. We are merely gathering such lethal high-level nuclear waste with no clue what to do with it.

Nuclear Energy Cannot Substitute Fossil Fuel

Even if the high-level nuclear waste disposal problem were disregarded, The U.S., which already has 100 nuclear power plants, would need to double its nuclear power generation to substitute for oil and natural gas. To substitute coal, which generates most of the country's electricity, 250 more plants would be needed. In other words, to provide the U.S. with its energy needs solely by nuclear power, more than 300 more plants would need to be built.

That total does not even include the energy used for transportation, which would require the current nuclear energy capacity to increase by 500 percent. Moreover, if the oil used by seven hundred million cars around the world were substituted with nuclear generated electricity, severe technological and economic problems would arise. Even if only half of this energy were substituted with nuclear energy, uranium fuels will be depleted in under thirty years, which would mean that plutonium for fast-breeder reactors or nuclear fusion reactors would need to be used. As was shown in the previous section, these technologies are incomplete and the possibilities are unsure. If an accident occurs due to an urgent operation under insufficient security, humanity will have to pay a great price from radiation and live in regret of its use for centuries to come. Therefore, we must stop thinking of nuclear energy as an alternative energy to fossil fuels.

Global Warming and Nuclear Power

Nuclear power is pretending to be a clean energy, while everyone has blamed CO_2 as the main culprit of global warming. However, we must rethink nuclear power and judge rationally. Previously, our simple experiment showed that CO_2 might not be the main cause of global warming. If CO_2 is not so important to global warming, then we must shift the focus from CO_2 to the heat generated by humans. And this heat is not only generated from our daily energy producing activities, but also from nuclear power generators. Nuclear power must be reviewed not only because of its danger, but also because of its huge use of coolants, which must also be a cause of global warming. In that context, let us look more closely at the reality of nuclear power generation.

First, a 1,000MW nuclear reactor discharges four million tons of coolants (water) daily and this causes the temperature of all of this water to increase about 13 degrees Fahrenheit, this heat has to go somewhere. Secondly, because nuclear power requires so much coolant, plants are usually installed near the coast and seawater is used. In contrast to fossil fuels, where the heat is released into the atmosphere, heat generated from nuclear power is stored in the ocean and is slowly released into the atmosphere.

Third, because of the large size of the oceans, it seems as though there is no influence from this heat. However, due to the nature of water, this hot water stays in the upper part of the ocean and influences only from ten to one hundred meters below the surface. The convection currents occur only near the surface and there is no temperature change in the deep ocean. The heated water travels due to ocean current and eventually affects the Arctic Ocean. This might be an important reason why the temperature increases rapidly in oceans near East Asia and Europe.

Nuclear power has serious problems. The problem is connected to the energy itself and also Earth. We must think hard about whether we should continue to use nuclear energy.

5. QUEST FOR A NEW CIVILIZATION

No fossil fuels. No wind and solar energy. No nuclear energy. It seems that there is nothing our civilization can depend on. Then, is there no future? There will be a future worth living in if we actively prepare and plan, and if everyone works hard. A future that everyone can handle and accept can be obtained if we change ourselves. We must ask ourselves what should be done? And then, we must do it every day.

Actions Required

The actions required are quite simple. The priority is to solve the imminent energy problem. However, it seems that the problem cannot be overcome by mere time-wasting efforts. For example, we cannot solve the fundamental problem with only energy saving campaigns. This effort must be done in larger frames than just individual ones. These include revolutions in social systems and the reformation of culture. In a broader sense, this problem must be approached by the entire planet, rather than by one society or country. Actions that are required can be summarized into two general points.

First are energy measures. Until the industrial revolution, our civilizations were maintained with limited resources. When threatened by these limited amounts, we turned our head to the 'unlimited' energy sources. But of course, the unlimited energy resources are not unlimited. All sources of energy are limited. We have depended on fuel that is impractical and dangerous.

Perhaps we should rethink about 'unlimited' energy to something that is 'endless'. One way that something is 'endless' is if it circulates. Circulating energy? Is such a thing possible? If circulating energy is all we have, we must seriously consider it.

The second important point is to change our culture. We have previously looked at the end of the fossil fuel civilization. This has always been inevitable because fossil fuels are limited. However, the fact that fossil fuels are limited is not the only reason for the end of the fossil fuel civilization. The social structure based on capitalism had a great effect. Ranging from daily lives to the social structure, our culture played a part in accelerating the end of fossil fuels and is still darkening our future. Unless our mindsets change, even obtaining circulating energy is useless. Circulating energy is effective only when we have the appropriate mindsets. It is like riding a bicycle. If the circulating energy is the right foot, the culture change is the left foot. Only when both feet pedal together can we ride towards happiness. Now, let us look at how circulating energy might be possible.

Circulating Energy

Civilizations existed before the fossil fuel civilization. Perhaps we can look at them to find appropriate energies. Do circulating energies exist? These thoughts lead directly to plants, the only energy resource that circulates. The main source of energy for much of history, for places that have them, were trees. Trees may lead to some answers

Peruvian historian, Garcilaso de la Vega, stated this poem about the Sun:

> I do good to all the world. I give them my light, and brightness that they may see and go about their

business; I warm them when they are cold; and I grow their pastures and crops, and bring fruit to their trees, and multiply their flocks. I bring rain and calm weather in turn, and I take care to go round the world once a day to observe the wants that exist in the world and to fill and supply them as the sustainer and benefactor of men.

Not only men, but organisms live thanks to sunlight. Most of the energy needed for life comes from the Sun. For us to utilize this energy, it must be taken from plants. When sunlight falls on Earth, plants take it to form leaves, stems, flowers, seeds, and roots. Animals eat animals that eat plants. This is the circulation of the ecosystem. Humans have used these plant parts as food and fiber to create clothes and paper. They are also used as medical herbs and material for building houses. Heat energy can also be provided by firewood. Plants provided us with everything we needed in our daily lives.

This is not simply an exercise in reminiscing about the past. It is an effort to find a solution from past experiences. Past experiences are not just memories but help form practical wisdom for future actions. Therefore, we are starting from the time when we used to depend on plants, before there were fossil fuels. And in the process, we should try to think differently about trees, a representative of plants. In reality, faster growing plants such as bamboo and hemp are even more effective for most energy and industrial uses.

When most people think of 'trees' as an energy source, they only think of firewood. This is because we have dealt with trees only in such a way for a long time. For many people with this viewpoint, talking about a tree-circulation civilization creates repulsion. They worry about the limited amount of energy, natural hazards and ecosystem destruction from cutting down trees for firewood. This sort of approach should be abandoned. It *must* be abandoned. Tree energy is not the sort of energy we simply use and throw away. To live with trees as a circulating energy resource, we must set our lives to circulate in the same ways.

Indiscriminately cutting down trees to get firewood shouldn't be done and can't be done for long. Rapid depletion from cutting down raw trees for firewood is a consequential result. If civilization depended solely on firewood, it would last less than a year. Some civilizations have already experienced this. There is no reason to repeat such foolish action. In the fossil fuel civilization, trees were considered only in terms of environmental issues and not as a practical source for the necessities of life. This is because we were able to live an affluent life without it. But, as the fossil fuel civilization comes to an end, we will only have nature's circulating energy left. We must look again at trees.

Plants are by far the most efficient at absorbing

and storing solar energy. They grow anywhere if there is soil and water. They are exponentially more efficient than solar-cell panels that utilize only about five percent of solar energy hitting it. They also require energy to produce; they cannot just grow a solar panel. If we can apply our advanced technology to plants that are able to absorb and store ten times the amount of solar cells, it would be efficient in terms of energy, resources, the environment and land usage.

The Start of Tree Circulation Civilization

Since the beginning of human history, trees and forests have an inseparable relationship with humans. From the primitive ages up to the industrial revolution, humans earned their household items from the forests. However, there is no record of systematic management of the forests. They simply used what they needed and took no more. For example, people took fruits for food, different materials for farming and building houses from nearby mountains or hunted wild animals for means of living. In today's world, this simplicity may not be possible, but proper management could be the solution. What would happen if we stopped cutting down trees for unsystematic consumption and started towards a circulating civilization with trees?

First, let us look at the wood resource. If we plant new trees as we cut the wood, it takes about thirty to fifty years for a new tree to be ready to harvest. If the amount of wood used does not exceed the amount of trees growing, trees can be used as a sustainable energy source. If this proceeds, trees can be used for daily life. New plant to harvest time is reduced to as little as a year for fast growing plants like bamboo and hemp. While these plants do not absorb nearly the same level of energy as the lifecycle of a tree, where they can be used in place of cutting trees would mean saving the harvesting of trees to limited and specific purposes.

The application possibilities are endless, but include the following:

First, clothing can be perfectly solved with plants. Most of the clothes we wear are made through petro-chemistry. Would we be living naked in a tree-circulation civilization? Never! We made our clothes from natural materials before fossil fuels. Yarning fibers from plants are the easiest ways to obtain material from plants. Hemp is one of the most useful plants, with an estimated 50,000 industrial uses, including fibers for jeans and ropes. There are likely many types of vegetable fibers that we have never even considered using before, but will be discovered in a tree-circulating culture.

Food is abundant in nature, and basically all healthy foods come from plants. Excessive meat consumption takes a lot of energy, or oil, and is unnecessary. Animals will grow naturally from feeding on natural grass in this tree culture. Humans would eat a more natural and healthy amount of meat. Housing and shelter is also abundant in nature. Trees are the oldest and the best material for construction. Wooden houses are strong, long-lasting and healthy. New methods of creating concretes out of plant fibers are also highly effective and efficient.

Life in nature may not be perceived to be as comfortable or rich as life in fossil fuel civilization, but perhaps it would be happier. We have become accustomed to the excesses and opulence that oil has provided. What exactly is a rich and comfortable life? It means plenty of food and clothing, living in air-conditioned or heated houses, and that physical movement is easy but not necessary.

Whether the oil-based life is 'good' is not an easy question to answer. However, such a life has led to social inequality and major mental and physical side effects. It might seem good temporarily, but in the long-term, it has given birth to undesired and harmful results. Synthetic fibers, synthetic drugs, genetically modified fast foods, and over-consumption are not helpful to our health, society, or future. We are too accustomed to these things and living without them seems scary. We are fat and complacent with the way things are, but our fear of change just might kill us.

The fossil fuel civilization is reaching its end and people must free themselves from such obsessions. A tree-circulation civilization will change our daily lives; it is a way back to a natural and healthy life. The necessity of this change is based upon valid reasons, not some idealized fantasy about natural living.

Now, we must think again about energy. Can we really solve the energy problem with trees? Perhaps, if we change our mindset. For example, consider wooden houses. When a house's lifespan ends, that wood should be used again, or used as firewood. Unlike using raw trees as firewood as we did in the past, we would be using used-wood as firewood. We should no longer treat trees as mere firewood. Plants should be seen as the only resource left to humans and a gift from God; the basis of life and massive storage of solar energy. The only question is how efficiently can we use the stored energy?

As stated earlier, the only way we can look forward to a bright future is by changing ourselves. This change is even more important than simply using a circulating energy system. It is necessary to ensure our survival. How do we change our awareness?

Change of Awareness

Humans have ruled the earth ever since they found fire. This means more than simply sitting at the top of the food chain. Humans were able to avoid danger and threats from the wild and to live a stable life. This allowed them to infinitely expand their awareness. The discovery of fossil fuels gave humans even more possibilities and rapid development was made possible. Fossil fuels were the basis of exponential expansion of human awareness. New materials that did not exist in nature were made and modern civilization was built on fossil fuel. A switch of energy allowed such change.

The current situation is not much different. We are facing the end of the fossil fuel era. Therefore, a switch of energy is inevitable. This will also lead to another change in awareness. Like how fossil fuels expanded our awareness rapidly, its depletion will require some kind of awareness change also. Humans, accustomed to a fossil fuel civilization, will change. Without change, our future will be bleak. Without radical change, the tree circulation civilization will also run into a fast end. We cannot treat plant resources the same as we treated oil.

Switching energy means changing life. All this is only possible when our awareness and mindset changes. Changing the large frame of society is difficult. However, if we are ready for the new life with a tree circulation civilization, our future is bright. It is best to plan ahead before the problem occurs and everyone on Earth must work together to make it possible.

We learned many things by using fossil fuels. Physics and chemistry have progressed to where we can observe galaxies a billion light years away and research quarks and neutrinos. Also, the synergism of science, religion and philosophy allowed us to think about human life in a much broader scope. The two hundred years of fossil fuel civilization has left us with vast knowledge, technology and wisdom. We must use what we have learned to prepare for life in a tree circulation civilization. The new civilization cannot have wars, selfishness or greed, in order and provide a life that can coexist with other organisms.

The change of awareness starts here. If we can use fossil fuels for fifty more years based on the current consumption rate, we should start reducing the fuel consumption to maximize the time, and meanwhile, prepare for the tree circulation civilization. If we reduce fuel consumption by half during the next ten years, we can use fossil fuels for another eighty years. If we do this three times for the next thirty years, we can still use fossil fuels for another three hundred years.

There are two obvious problems with this idea. First, how can consumption be reduced by half when consumption levels are still increasing, and the other is that the end of fossil fuels will still come to an end after three hundred years anyway. It is important to understand that this idea is not merely trying to extend the fossil fuel civilization. We are hoping to smoothly and calmly settle into a tree circulation civilization while extending the current civilization. If humanity is fully aware of this and adapts to the necessary changes, it is possible. If not, the fossil fuel civilization will still end, except it will feel sudden and chaotic. The required changes in awareness and mindset come from answering these and other important questions.

- Does economic growth have to be positive?

- Is there any problem from population increase?

- If resources get scarce, should we support people who have lived unhealthy lifestyles?

These questions are tightly related to the energy problem. First, energy and labor is needed for economic growth. For positive growth, we must manufacture and consume things and the labor population must not decrease. Consequently, the problem of an aging and sick population has emerged all over the world. These are the results of the immutable law of "economic growth". Are these ideas right? We must turn these ideas upside down to face hope. Isn't negative economic growth okay? Shouldn't the population decrease? Can't we make a condition where old people can live more healthy and independently?

We must not forget the premise. We are going to face a situation where growth will be impossible. The economy cannot always grow; nothing in nature has constant growth except cancer, and that growth ends when the constant growth kills the host. Let us think why economic growth is needed. Economic growth means more energy consumption. Putting it simply, consuming fuel energy in different ways is called economic growth. Look at large hotels as a good example. Why such big accommodations? Largely, this is due to the tourism industry and the fact that many people are trying to escape their unhappy daily lives. This business is obviously linked to fuel resources used to travel around. When the economy is down, energy is expensive, tourism is down. From an energy point of view, the market economy is an economic structure that competitively consumes fuel. Building roads, bridges and buildings are impossible without fossil fuel. What happens if we eliminate the fuel factor? Everything that has been built will fall like sand. Are things like economic growth and the global economy possible when the premise is falling apart?

The energy crisis will test the social system. The industrial structure will be adjusted and the industries that are important today could be thrown out tomorrow. We should not artificially prop up what will naturally fall. It will be difficult for heavy energy industries, such as shipbuilding and car industries, to bear with the energy crisis. Many people will become unemployed and feel ill prepared for the new economic environment. Unless, they are given other avenues of employment that are in harmony with the tree-circulating civilization. The change in social structure will follow. The current system cannot endure. There are limits for a society that promotes consumption.

It seems that there is no way other than to lower the growth goal in each area. This will reduce the amount of fuel consumption and help us to settle into the tree circulation civilization. Effort in each area that follows the change in awareness and mindset will help us easily move from fossil fuel civilization to tree circulation civilization.
The next section will briefly look at efforts that should be done in different areas.

Economy: Shift From Global Economy to Regional Economy

The current economy is based on mass production and distribution of goods among countries. Economy is the most globalized field in the fossil fuel civilization. For example, most successful countries fall into one of these categories: large landmass with a good climate focused on agriculture, human resources and advanced technology focused on industry, and countries with many historic sites focused on tourism. However, as previously discussed, if petroleum gets depleted, there will be no fertilizers, pesticides or tractors to help farming. Industries such as auto, steel, shipbuilding, chemical, electricity, telecommunications, will all stop without fossil fuels. The tourism industry obviously cannot be maintained in that situation. What should we do now?

The global economy must be ready to turn back into regional economies. Since sudden change is too much of an impact, we must reduce the scale of heavy industries by half for each decade. Also, to reduce unemployment, the machine-oriented manufacturing should be changed to manpower-oriented manufacturing. A new tree industry must be nurtured and substitute petrochemical products to plant-based products. Researching and shifting production of many of today's necessary products to be made from plants instead of oil would create many jobs. Construction of roads, ports, airports, railroads and any other unnecessary large-scale development project should be cut.

Population: Movement Towards Optimum Population

What would be the optimal population that would allow us to coexist with other organisms? While the human population is rapidly increasing, the population of other creatures is rapidly decreasing. In fact, scientists consider us to be living in a major extinction period, one not caused by asteroids, but by humans. The current human population is about seven billion. Can the entire population live together when the petroleum gets depleted? No one likes to think of this problem. However, like every organism, there is a natural function that controls populations when they reach a certain resistance, humans are no different. Since it is hard to determine the optimum population regarding the entire population on Earth, it is best to determine in accord with each country. These actions are only possible when everyone understands why these efforts are needed.

Military: Awareness of Unnecessary Energy

It is hard to exactly determine the amount of energy consumed in militaries all over the world, but it accounts for a large portion of the entire energy consumption. In the U.S., it is by far the largest consumer of oil of any government function. If we assume that 20% to 30% of global energy is consumed by militaries, it is worth taking a close look at what this energy means to humanity.

In the past, the main purpose of a military was to fight a war to take away territory, food and labor. It was an easy solution to the increasing population in each country. However, with the use of fossil fuels, the necessities of life were easily solved and the purpose of a military has changed. Although countries say that their military serves to protect from the invasion or threat of others, in reality, it is to make and keep their profit. It is a method to survive in the economic warfare. All countries are based on an energy economy, and the military is preparation for access to oil. Current and future wars will be for oil and the conflicts will get worse once oil production decreases. There is a decent chance that they will escalate to using nuclear weapons, at which point it would be even harder to create a nice future. Let us think. Is it necessary to spend large amounts of oil by maintaining a military force, in order to gain access to oil, which is depleting? Doesn't this seem ridiculous and meaningless? If we consider the danger this poses to all of humanity, we must first reduce the energy consumed in maintaining military.

Is it possible to eliminate all militaries? No one would think so. If everyone understands the problem and knows which way to go, the leader of each country will pick up the public opinion and discuss the issue in the U.N. A solution will appear if each country understands it is in their best interest. Halving the fossil fuel consumption can be easily achieved if we reduce the energy consumed in military powers.

Culture: Breakaway from Culture of Consumption

In the past, the basis culture was made by survival needs. But, the culture in this fossil fuel civilization is different. The necessities of life were easy to obtain. The things needed for survival were easily resolved. Thanks to fossil fuels, people naturally took interest in other things and started consuming fossil fuel energy competitively. The culture of consumption was developed. Is a culture of consumption bad? Not inherently, but when the natural structure is not built to handle it, the results will be disastrous.

There is a need to reduce our culture in scale. For

example, rather than build large elaborate theatres, we should use outdoor stages. For sports, we should only use the stadiums that are already built, rather than demolish and rebuild it every twenty years. If we enjoy tourism and leisure activities efficiently, we can save a lot of energy.

Energy: Petroleum Product Waste and Trees

Heat energy is very important for humans and has a close relationship with housing. In the fossil fuel civilization we did not need to worry about heating because of the abundant amount of petroleum. But in the tree circulation civilization heating might be the most important. We must think about how to efficiently use the limited amount of trees.

During the switch from civilizations, the waste from petroleum products should not be buried or burned. Too much energy is used if we are to collect, store, transport, and reprocess the waste; we must find a way to use the waste as fuel in each process. But can we simply burn the waste from petroleum products? You don't have to be an environmentalist to know there are problems from burning those products. It emits carbon monoxide, nitrogen oxides, sulfur oxides and dioxin, which are all harmful to humans and other organisms. The toxic gases, excluding dioxin, can be neutralized in the natural environment. However, dioxins remain semi-permanently if it is not heated up to 700 to 800°C. It accumulates on the planet. Therefore, many countries have laws preventing direct burning of petroleum products. It is only allowed to be burned in large incinerators managed by countries. It seems we can find a new method. If large incinerator technology is applied to create small boilers and incinerators (currently being operated in designated areas), we can solve the dioxin, waste and housing energy problem all at once, by installing them in individual houses to use as heaters.

These same stoves can use energy by burning the wastes using the heat from burning old wood. Household incinerators can be easily made with the current level of technology. Then, the household heaters and boilers will also have an incinerating function. By applying these to floor heating systems and pre-existing chimneys, comfortable household heating systems can be created, and we can easily transition to a tree circulation civilization.

Science Education: Proper Understanding of Our Life

This shift in civilization cannot be accomplished in

one generation; therefore, the next generation must accurately understand the current situation in order to continuously carry forward the work. Education is vital to accomplish this task.

In the fossil fuel civilization, competition between individuals for 'the good life' is more valued than the life of communities. Therefore, education is currently more focused on teaching how to get ahead, take advantage of others, earn more profit and gain higher status. Once the fossil fuel civilization is gone, we are going to require a more communal way of thinking than individualistic. Therefore, education must be reformed to teach people the basic wisdom needed for life, instead of teaching to create a competitive society. For tree industries to substitute fossil fuel industries, we must learn and research how to associate past survival methods and tools with current scientific technology. Tree industries will increase in efficiency. The way we educate will change based on the movement in civilization.

A tree circulation civilization does not mean that we will only use trees. Like in the Iron and Bronze Ages, we can carefully use various metals with plants. However, the amount of metals will be limited. Mass consumption should be controlled and metal should be reused. If we minimize the use of automobiles, airplanes and ships and change the fuel energy of power generators to biomass and plant-energy power generators, we do not necessarily have to live a completely manual life.

If we convert the education focused on oil science to tree science, a lot of discontinued past tools can be redeveloped. For example, in the past, people used A-frame carriers to lift heavy objects rather than say a forklift. Although it is hard to lift an object weighing 150 pounds without an A-frame carrier, it was easily possible to lift two bags of rice with this device. This is just one example of how we have developed dependences on powered machinery when the same job could be accomplished without it with some ingenuity. With the help of modern science, these old devices can further evolve to make them more efficient. Like how many Swiss watches are made by hand, small equipment can be made in small factories by manpower. This will make technical training important, and skilled workers will again replace robots. Young people can plan their future based on their talents, and break free from the current competitive education.

Information, Communication: Informing About the Crisis and Alternatives

The climax of fossil fuel civilization has been the information and communication revolution. This not only brought new knowledge, but it also allows us to instantly share our thoughts with everyone, receive feedback, and act upon it together. Most people assert that humans should coexist with all other organisms, very few would disagree. How many will act, though? Would you simply sit back and watch the animals and plants go extinct due to the possible nuclear war that will erupt because of fossil fuels?

With the help of information technology, many will apprehend and judge the situation accurately and act accordingly. If there are enough people who realize that crisis is upon us, they will communicate and prepare a solution. This cannot be realized with just the power of a small group. Shared understanding and changes in ways of living, will power the changes that are necessary to make our future worth living in.